Deep Space Discovery

THE SUN AND OTHER STARS

GAIL TERP

BLACK RABBIT BOOKS

Bolt is published by Black Rabbit Books
P.O. Box 3263, Mankato, Minnesota, 56002.
www.blackrabbitbooks.com
Copyright © 2019 Black Rabbit Books

Marysa Storm, editor; Grant Gould, designer;
Omay Ayres, photo researcher

Library of Congress Cataloging-in-Publication Data
Names: Terp, Gail, 1951- author.
Title: The sun and other stars / by Gail Terp.
Description: Mankato, Minnesota : Black Rabbit Books, [2019] | Series:
Bolt. Deep space discovery | Audience: Ages 9-12. | Audience: Grades 4
to 6. | Includes bibliographical references and index.
Identifiers: LCCN 2017019821 (print) | LCCN 2017036081 (ebook) |
ISBN 9781680725407 (ebook) | ISBN 9781680724240 (library binding) |
ISBN 9781680727180 (paperback)
Subjects: LCSH: Sun–Juvenile literature. | Stars–Juvenile literature.
Classification: LCC QB521.5 (ebook) | LCC QB521.5 .T465 2019 (print) |
DDC 523.7–dc23
LC record available at https://lccn.loc.gov/2017019821

Printed in China. 3/18

Image Credits
Alamy: Jurgen Ziewe, 9 (bkgd);
jpl.nasa.gov/: NASA, 14 (black dwarf);
nasa.gov: NASA, 28–29; NASA/ESA, 14
(star); science.nasa.gov: NASA, 16–17; Science
Source: Mikkel Juul Jensen, 10–11; Paul Wootton,
24–25; Science Picture Co, 22–23; Shutterstock: alex-
aldo, 24; Anton Balazh, 6; Azuzl, 20–21; Dja65, 26–27;
GiroScience, 15 (red giant); In-Finity, 27 (constellations),
31; Jurik Peter, 3, 32; Maria Starovoytova, 12; NPeter, 18;
sciencepics, 15 (white dwarf); suns07butterfly, 27 (bkgd);
TeddyGraphics, 9 (Jupiter); Twin Design, Cover (sun), 1;
Yuriy Mazur, 14–15 (bkgd), 21 (bkgd); solarsystem.nasa.
gov: NASA, 4–5; www.noa.ac.jp, National Astronomi-
cal Observatory of Japan, Cover (satellite)
Every effort has been made to contact copyright
holders for material reproduced in this book.
Any omissions will be rectified in subse-
quent printings if notice is given
to the publisher.

CONTENTS

CHAPTER 1

The

The sun is Earth's closest star. It gives Earth light and keeps the planet warm. But what if the sun suddenly disappeared? A lot would change. It would always be night on Earth. Without sunlight, plants could not make food. Most plants would die. And the planet would get very, very cold.

5

Deep Freeze

In the first week without the sun, Earth would get cold. Temperatures would be around 32 degrees Fahrenheit (0 degrees Celsius). After one year, it would be much worse. Oceans would freeze over. Only tiny creatures could live beneath the ice. It's clear Earth needs the sun.

The sun is about 4.6 billion years old.

The Solar System

The sun is at the center of a solar system. Planets circle around the sun. The sun's **gravity** pulls Earth and seven other planets to it. The sun also holds dwarf planets, such as Pluto, in **orbit**.

The sun isn't the only star with a solar system. Other stars are part of solar systems too.

Earth's solar system has at least
149 moons. They circle dwarf and regular planets.
The planets' gravity keeps moons in their orbits.

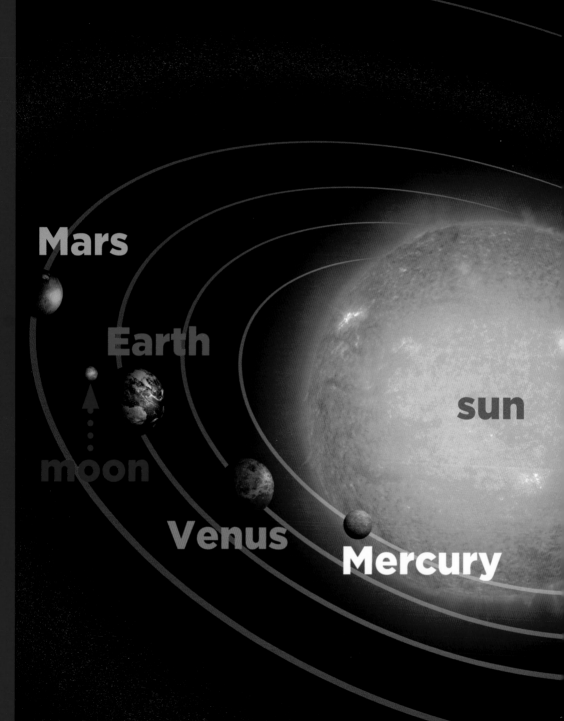

Mars

Earth

moon

Venus

Mercury

sun

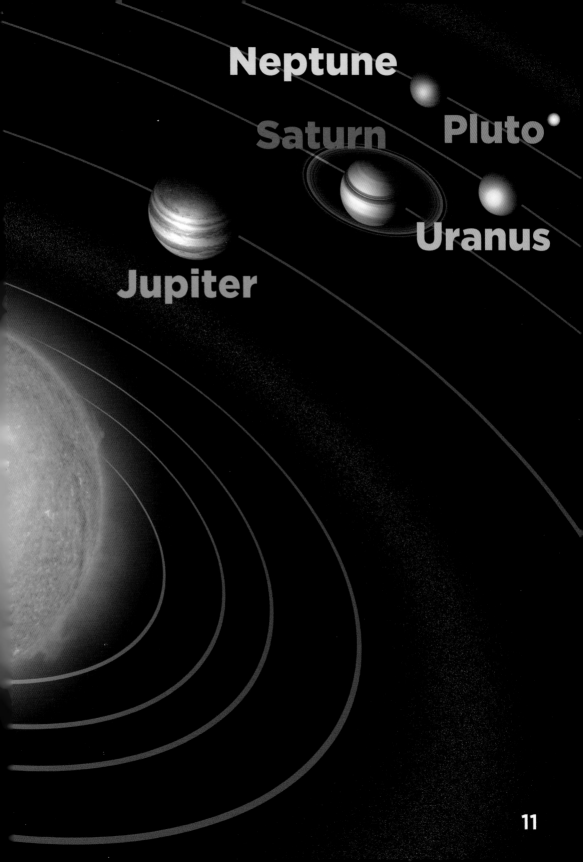

Neptune

Pluto

Saturn

Uranus

Jupiter

CHAPTER 2

STARS

Earth's sun started out as a cloud of gas and dust. That is how all stars begin. Gravity then pulled the cloud of gas and dust into a clump. Heat in the center of the clump grew. In time, the clump's main gas changed. It became a different gas. That change created energy. The energy makes the sun shine.

A star's color depends on its temperature. Blue stars are the hottest. Red stars are cooler.

THE LIFE OF A STAR

Clouds of gas and dust, called nebulas, form protostars. Protostars become stars. Once stars form, they go through several stages.

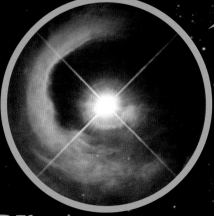

star

The gas and dust of a protostar becomes hot enough to burn. The main gas changes, and a star begins to shine.

black dwarf

The star slowly cools. It becomes too faint to detect.

red giant

After millions to billions of years, most of the star's **fuel** has burned. The star's outer layers expand. It glows red.

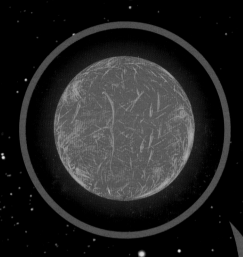

white dwarf

A star becomes a white dwarf when all its fuel has burned. Only the core remains.

Pairs, Clusters, and Galaxies

Stars can have partners. Two stars that orbit each other are part of a **binary** system. Scientists believe binary systems develop when a star first begins to form. The clump of dust and gas splits into two. Then two stars form.

Large groups of stars can also be found in space. Scientists call them star **clusters**. Some clusters have a few hundred stars. Other clusters have millions. Gravity holds the cluster together. Galaxies are giant groups of stars. They can have trillions of stars.

The sun's solar system is part of the Milky Way galaxy. The galaxy has 100 to 400 million stars in it.

light-year

A light-year measures distance. It's how far light travels in one year. One light-year is about 5.9 trillion miles (9.5 trillion kilometers).

Famous Stars of the Night Sky

Earth's sun is just one of many famous stars. Polaris is another well-known star. Many people call it the North Star. Sirius is the sky's brightest star. It's more than 25 times brighter than the sun. The star is 8.6 light-years from Earth. Alpha Centauri is the third brightest star. Alpha Centauri is even closer to Earth. It's 4.3 light-years away.

CHAPTER 3

A CLOSER LOOK
at the Sun

Like other stars, the sun is made of gases. These gases are extremely hot. The sun's surface is 10,000 degrees Fahrenheit (5,538 degrees C). Its center is a whopping 27 million degrees Fahrenheit (15 million degrees C).

Sizing Them Up

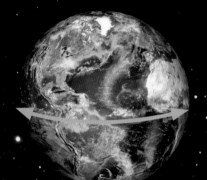

Earth
diameter
about 7,926 miles
(12,756 km) at equator

sun
diameter
864,337 miles
(1,391,016 km)

PARTS OF THE SUN

The sun is more than just a plain ball of gas.

prominences
bright features seen above the sun's surface

core
where most of the fuel burns

photosphere
surface of the sun

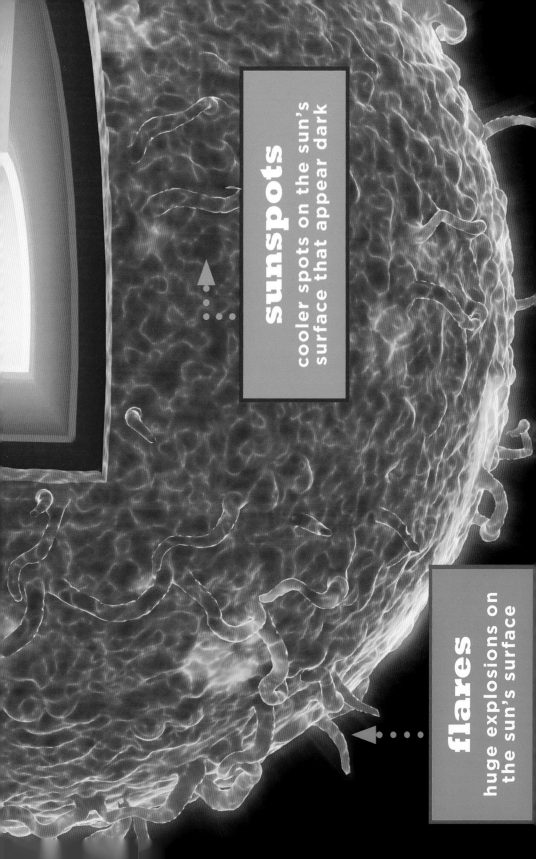

sunspots
cooler spots on the sun's surface that appear dark

flares
huge explosions on the sun's surface

sun

A Solar Eclipse

Sometimes Earth's moon passes between the sun and Earth. This event is called a solar eclipse. During an eclipse, the sun's light is blocked from Earth. It only lasts a few minutes. A total eclipse can be seen from somewhere on Earth every 18 months.

Solar Eclipse Diagram

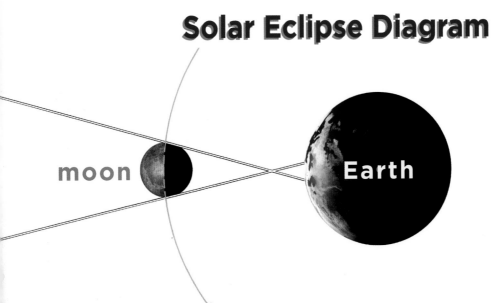

moon Earth

CHAPTER 4

LEARNING
More

People have studied stars for hundreds of years. To **ancient** people, the stars told stories. Sailors used the stars' positions to **navigate**. Stars' movements helped people make the first calendars too.

Constellations are named groups of stars that form shapes in the sky. Most constellations are part of ancient stories.

The Parker
Solar **Probe**
will go within
3.7 million miles
(6 million km)
of the sun.

Future Sun Explorations

Scientists still study the sun and stars today. They use telescopes and probes to do so. One current mission is Parker Solar Probe. This probe will orbit the sun 24 times. Scientists hope to gather more data about the star.

Each year, scientists discover more about stars. Maybe someday, they will find another star that supports life.

ancient (AYN-shunt)—from a time long ago

binary (BAHY-ner-ee)—something made of two things or parts

cluster (KLUHS-ter)—a group of things that are close together

equator (ih-KWEY-ter)—an imaginary circle around Earth that is equally distant from the North Pole and the South Pole

fuel (FEYUL)—a material, such as coal, oil, or gas, that is burned to produce heat or power

gravity (GRAV-i-tee)—the natural force that pulls physical things toward each other

navigate (NAV-i-geyt)—to find the way to get to a place when you are traveling

orbit (AWR-bit)—the path taken by one body circling around another body

probe (PRHOB)—a device used to collect information from outer space and send it back to Earth

BOOKS

Hudak, Heather C. *The Sun.* Exploring Our Universe. Minneapolis: Checkerboard Library, 2017.

Ponka, Katherine. *Math On the Sun.* Math in Space. New York: Gareth Stevens Publishing, 2017.

Terp, Gail. *The Milky Way and Other Galaxies.* Deep Space Discovery. Mankato, MN: Black Rabbit Books, 2019.

WEBSITES

Sun
kids.nationalgeographic.com/explore/space/sun/#sun.jpg

Sun
spaceplace.nasa.gov/menu/sun/

The Sun
www.ducksters.com/science/sun.php

INDEX